UPDOG

WORLD'S GROSSEST

WORLD'S GROSSEST ANIMALS

Scott Nickel

Lerner Publications ◆ Minneapolis

Lerner Publications Company
An imprint of Lerner Publishing Group, Inc.
241 First Avenue North
Minneapolis, MN 55401 USA

For reading levels and more information, look up this title at www.lernerbooks.com.

Main body text set in ITC Franklin Gothic Std.
Typeface provided by Adobe Systems.

Designer: Mary Ross

Library of Congress Cataloging-in-Publication Data

Names: Nickel, Scott, author.
Title: World's grossest animals / Scott Nickel.
Description: Minneapolis : Lerner Publications, [2022] | Series: World's grossest (updog books) | Includes bibliographical references and index. | Audience: Ages 8–11 | Audience: Grades K–1 | Summary: "Explore the disgusting animal kingdom, from horrible hagfish to vulgar vultures. Gross photos and a disgusting diagram give readers a close-up look at nature's most sickening creatures"— Provided by publisher.
Identifiers: LCCN 2020003373 (print) | LCCN 2020003374 (ebook) | ISBN 9781728403960 (library binding) | ISBN 9781728418926 (ebook)
Subjects: LCSH: Animals—Miscellanea—Juvenile literature.
Classification: LCC QL49 .N52 2022 (print) | LCC QL49 (ebook) | DDC 590—dc23

LC record available at https://lccn.loc.gov/2020003373
LC ebook record available at https://lccn.loc.gov/2020003374

Manufactured in the United States of America
1-48454-48968-5/5/2021

TABLE OF CONTENTS

Creepy Creatures 4

Sea Monsters 10

Diets and Defenses 16

Glossary 22

Check It Out! 23

Index 24

CREEPY CREATURES

Get ready for some of the grossest, smelliest, and slimiest animals in the world. Read on, if you dare!

Also called a sand puppy, the naked mole rat lives underground.

The star-nosed mole can't see.
Its fleshy nose helps it find food.

The hyena’s strong jaws can crush bones.

How did the hoatzin get the name stinkbird? It smells like cow poop!

The snot otter is a slimy amphibian.

AMPHIBIAN:
an animal that can live in water and on land

UP NEXT!
SLIMY SEA CREATURES.

SEA MONSTERS

The goblin shark stretches its jaw to sweep up prey.

PREY:
an animal that other animals hunt for food

The giant isopod is a crustacean. It can grow to over 30 inches (76 cm) long.

CRUSTACEAN:
an animal with a hard covering that lives in water

DISGUSTING DIAGRAM!

The sea lamprey has rows of sharp teeth.

The lamprey attaches to a fish. Its mouth acts like a vacuum, sucking out blood.

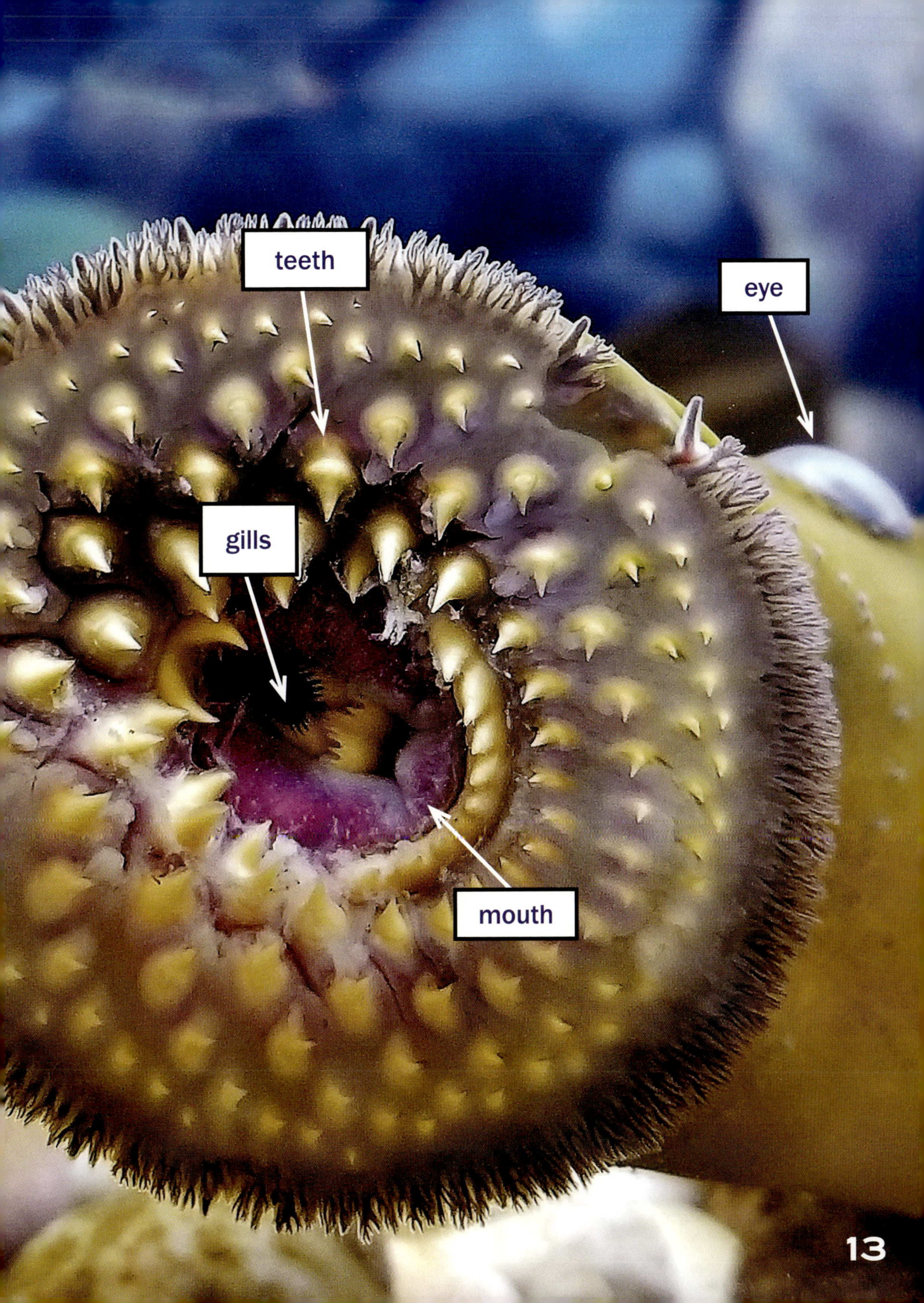
teeth
eye
gills
mouth

WHICH IS GROSSER?

Sea Cucumber vs. Hagfish

The sea cucumber is an invertebrate. When scared, it sheds its skin.

The hagfish is boneless and jawless. It produces tons of slime.

INVERTEBRATE: an animal that doesn't have a backbone

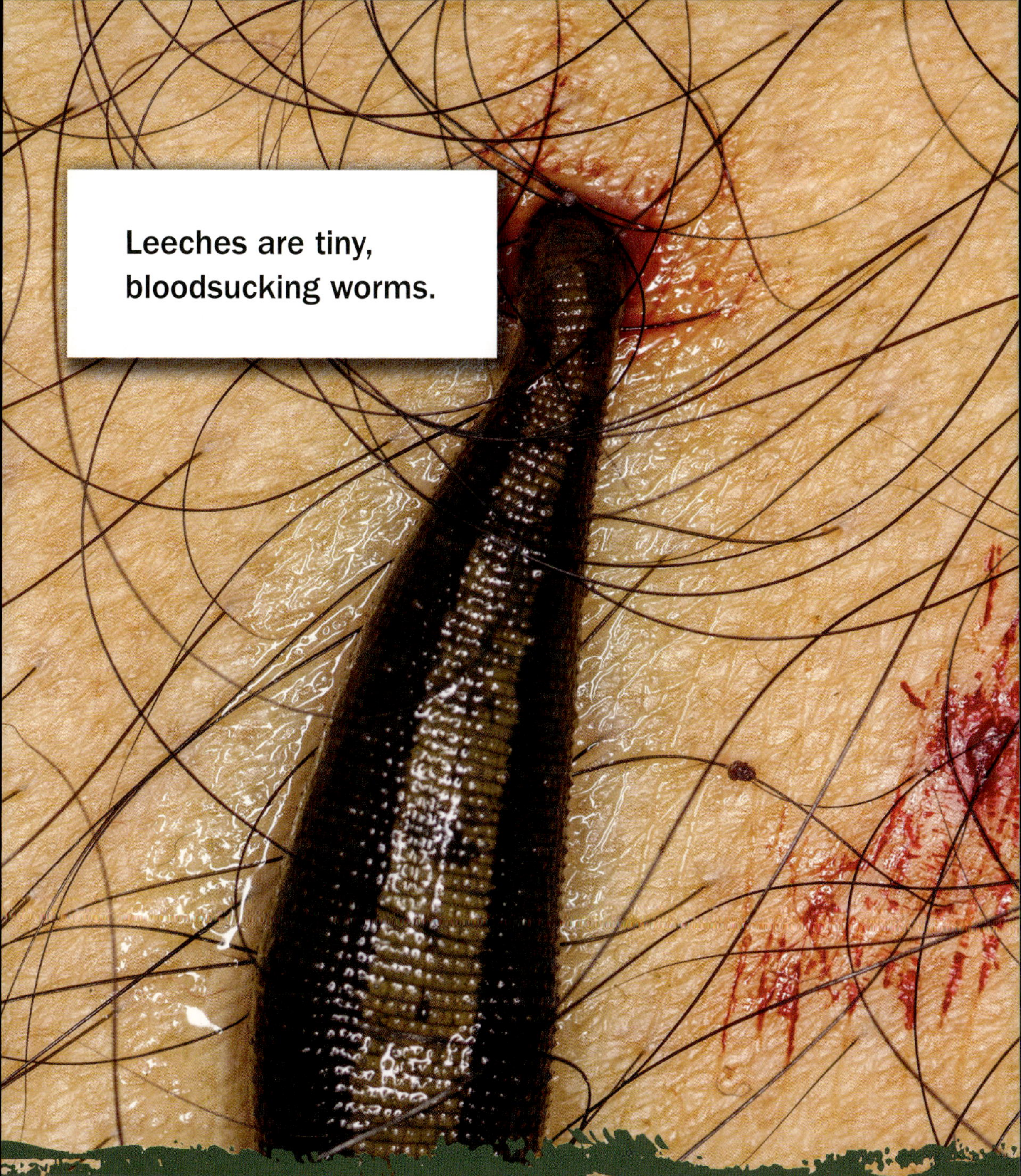

Leeches are tiny, bloodsucking worms.

UP NEXT!

ANIMALS ATTACK.

DIETS AND DEFENSES

The fulmar spits sticky stomach oil at predators.

PREDATOR:
an animal that eats other animals

Skunks aren't always stinky. But when they're scared, they squirt a smelly spray.

When the horned lizard is in danger, it shoots blood from its eyes.

The vampire bat has sharp teeth and drinks blood.

Vultures are scavengers. Their breath smells sour and rotten.

SCAVENGER:
an animal that finds dead animals to eat

Gross animals like this snot otter live all over the world!

Glossary

amphibian: an animal that can live in water and on land

crustacean: an animal with a hard covering that lives in water

invertebrate: an animal that doesn't have a backbone

predator: an animal that eats other animals

prey: an animal that other animals hunt for food

scavenger: an animal that finds dead animals to eat

Check It Out!

Braun, Eric. *Awesome, Disgusting, Unusual Facts about Animals*. Mankato, MN: Black Rabbit Books, 2019.

Mattern, Joanne. *Ugliest Animals*. Egremont, MA: Red Chair, 2020.

National Geographic Kids
https://kids.nationalgeographic.com/animals/

Nickel, Scott. *World's Grossest Creepy Crawlers*. Minneapolis: Lerner Publications, 2022.

SwitchZoo
https://switchzoo.com

ZooBorns
https://www.zooborns.com

Index

blood, 12, 14, 18–19

gross, 4, 21

jaw, 7, 10, 15

slime, 15
smell, 4, 8, 17, 20
stink, 8, 17

teeth, 12, 19

Photo Acknowledgments

Image credits: Bernard DUPONT/flickr (CC BY-SA 2.0), p. 4; Neil Bromhall/Shutterstock.com, p. 5; Stan Tekiela Author/Naturalist/Wildlife Photographer/Getty Images, p. 6; Picture by Tambako the Jaguar/Getty Images, p. 7; Morten Ross/Shutterstock.com, p. 8; Jay Ondreicka/Shutterstock.com, p. 9; Kelvin Aitken/VWPics/Alamy Stock Photo, p. 10; Corbari L./Museum national d'Histoire naturelle/Wikimedia Commons (CC BY 4.0), p. 11; Maria Dryfhout/Shutterstock.com, p. 13; Paulo Oliveira/Alamy Stock Photo, p. 14 (left); Laura Dts/Shutterstock.com, p. 14 (right); sydeen/iStock/Getty Images, p. 15; Andreas Trepte/Wikimedia Commons (CC BY-SA 2.5), p. 16; Karel Bock/Shutterstock.com, p. 17; J. Dullum/USFWS/flickr (CC BY 2.0), p. 18; Nicolas Reusens/Moment/Getty Images, p. 19; Mountaintreks/Shutterstock.com, p. 20; Jill Utrup/USFWS/flickr (CC BY 2.0), p. 21. Design elements: Saravudh Sudkhaw/Shutterstock.com; Green angel/Shutterstock.com.

Cover: Agnieszka Bacal/Shutterstock.com.